Understanding Machine Learning: Approaches, Algorithms, and Business Applications

ANDREI OPRISAN

DEDICATION

This book is dedicated to my wife, Jazmin C. Muñoz MD,
and my son Emil M. Oprisan.

CONTENTS

I. Introduction

Machine learning is a rapidly evolving technology that allows computers to learn from data without being explicitly programmed. It has the potential to transform businesses by enabling them to make more informed decisions based on data insights. This book is aimed at a business generalist audience with limited technical understanding who want to learn more about machine learning and its potential applications in business.

In this section, we will define machine learning and explain why it is essential for businesses to understand. We will also provide an overview of the book and what readers can expect to learn.

What is Machine Learning?

Machine learning involves training algorithms on data to identify patterns and make predictions based on that data. The algorithms "learn" by adjusting their behavior based on the data they are trained on and can then be used to make predictions on new, unseen data.

There are three main types of machine learning algorithms: supervised learning, unsupervised learning, and reinforcement learning. Supervised learning involves training an algorithm on a labeled dataset, where the correct output is known for each input. Unsupervised learning involves training an algorithm on an unlabeled dataset where the right result is unknown. Finally, reinforcement learning consists of an algorithm learning to take actions in an environment to maximize a reward.

Why is Machine Learning Important for Businesses?

Machine learning has become increasingly crucial for businesses seeking a competitive edge in today's data-driven economy. By training algorithms on data, companies can identify patterns and make predictions that can help them to improve efficiency and drive growth.

For example, a retailer might use machine learning to analyze customer data and predict which products will most likely be purchased by a particular customer. This can help the retailer tailor their marketing campaigns and inventory management to each customer's unique preferences, leading to increased sales and customer satisfaction.

In another example, a manufacturer might use machine learning to analyze sensor data from their equipment to predict when maintenance is needed. This can help the manufacturer to reduce downtime and improve efficiency, leading to cost savings and increased productivity.

Overview of this book

This book is designed to share an introduction to machine learning and its potential applications in business. It is organized into several sections, each providing a detailed overview of different aspects of machine learning. Section II will provide an overview of machine learning algorithms, including supervised, unsupervised, and reinforcement learning. We will explain how each algorithm works and provide examples of real-world applications.

Section III will discuss how to prepare data for machine learning, including data cleaning, data normalization, and feature engineering. We will explain why data preparation is essential and provide examples of best practices. Section IV will provide a practical guide to implementing machine learning algorithms in a business setting. We will discuss the steps in implementing machine learning, including data preparation, algorithm selection, and model evaluation. We will also highlight potential challenges and offer solutions.

In Section V, we will discuss various sectors of the economy that can benefit from machine learning, such as healthcare, finance, and retail. We will provide examples of real-world applications of machine learning in each sector and discuss how machine learning can improve operational efficiency and drive growth.

In conclusion, we will summarize the key takeaways from the book and provide final thoughts and recommendations for readers.

In the appendix, we will highlight current tolls and frameworks, a data preparation checklist, an algorithm selection guide, and a glossary of key terms.

Machine learning is a powerful tool businesses can use to gain a competitive edge in today's data-driven economy. By training algorithms on data, companies can identify patterns and make predictions that can help them to improve efficiency and drive growth. This book has provided an overview of the different types of machine learning algorithms and the real-world applications of each and offered a practical guide to implementing machine learning in a business setting. By following the advice and best practices outlined in this book, businesses can begin to unlock the full potential of machine learning and stay ahead of the competition.

Moreover, this book is designed for business people with limited technical understanding, providing them with a clear and concise experience of machine learning and its potential applications in business. In this book, we have used industry-leading articles and books to provide examples and offer practical guidance for implementing machine learning algorithms in different sectors of the economy or improving operational efficiency with ML.

By providing a clear understanding of the concepts and potential applications of machine learning, we hope to demystify the technology and encourage more businesses to explore its potential. While implementing machine learning in a business setting can be complex, it is well worth the investment. By leveraging the power of machine learning, businesses can gain a deeper understanding of their customers, optimize their operations, and drive growth.

In the next section, we will overview the different types of machine learning algorithms, including supervised, unsupervised, and reinforcement learning. Then, we will explain how each algorithm works and provide examples of real-world applications.

II. Understanding Machine Learning

Machine learning is a subfield of artificial intelligence that allows algorithms to learn and improve from data without being explicitly programmed. In recent years, it has become a critical tool for businesses that want to stay competitive in today's data-driven economy. Machine learning can help companies to make better decisions, optimize operations, and create new products and services that meet the changing needs of their customers. This section will comprehensively overview machine learning and its different types. We will also explain how each type works and provide examples of real-world applications.

Introduction to Machine Learning

Machine learning involves training algorithms to learn from data and make predictions based on that data. It differs from traditional programming, which consists of writing rules to solve a specific problem. With machine learning, the algorithm is trained on data and learns to find patterns on its own. This makes machine learning more flexible than traditional programming and allows it to adapt to changing data and circumstances. Machine learning has become increasingly crucial for businesses because of the growing importance of data in decision-making. With the explosion of data in recent years, companies can now access more than ever. However, it can be challenging to make sense of all that data and turn it into actionable insights. This is where machine learning comes in. By training algorithms on data, businesses can identify patterns and make predictions that can help them to improve efficiency, optimize operations and drive growth. Moreover, businesses can leverage machine learning to automate decision-making processes, which can help them save time and money. For example, a retailer can use machine learning to predict which products will likely be purchased by a particular customer and tailor their marketing campaigns to each customer's unique preferences. This can lead to increased sales and customer satisfaction.

Machine learning has rapidly evolved technology with many real-world applications in different sectors. The rise of machine learning has been facilitated by increased computational power, advances in algorithms, and the availability of large amounts of data. As such, it has become a vital tool for businesses that seek to gain a competitive edge in their industry.

Significance of Machine Learning for Businesses

Machine learning can enable businesses to make informed decisions based on data insights, gain a deeper understanding of their customers, and create new products and services that meet their needs. By training algorithms on data, businesses can identify patterns and make predictions that can help them to improve efficiency, optimize operations, and drive growth.

For instance, machine learning can be used in the financial sector for credit scoring, fraud detection, and economic forecasting. Healthcare can be used for medical diagnosis, drug discovery, and personalized medicine. In retail, it can be used for a product recommendation, supply chain optimization, and inventory management.

Machine learning can also be used in manufacturing, energy, transportation, and many other sectors. As a result, it has become a vital tool for businesses that want to stay competitive and adapt to changing market conditions.

Machine learning is a critical tool for businesses that seek to gain a competitive edge in today's data-driven economy. With the rise of big data, companies can now access more than ever. However, it can be challenging to make sense of all that data and turn it into actionable insights. Machine learning can help businesses unlock their data's full potential, make informed decisions, optimize operations, and drive growth.

Supervised Learning

Supervised learning is a type of machine learning that involves training an algorithm on a labeled dataset, where the correct output is known for each input. The algorithm learns to make predictions based on the labeled data and can then be used to make predictions on new, unseen data. Supervised learning is often used for tasks such as classification and regression.

Classification involves predicting a categorical output variable based on one or more input variables. For example, a retailer can use a supervised learning algorithm to predict whether a customer will likely buy a particular product based on their past purchase history, demographics, and other variables. In healthcare, a supervised learning algorithm can be trained to predict whether a patient has a particular disease based on their symptoms and medical history.

Regression involves predicting a continuous output variable based on one or more input variables. For example, a supervised learning algorithm can be used to predict the price of a house based on its location, size, and other variables. In finance, a supervised learning algorithm can be trained to anticipate the stock price of a particular company based on its historical performance.

Supervised learning algorithms are trained using a labeled dataset, which is typically divided into a training set and a testing set. The training set is used to prepare the algorithm, while the testing set is used to evaluate its performance. The goal is to train the algorithm to make accurate predictions on new, unseen data.

Supervised learning has many real-world applications in different sectors. For example, it can be used in marketing to predict which customers will likely churn and tailor marketing campaigns to retain them. In finance, it can be used to predict which customers are likely to default on their loans and adjust the loan terms accordingly.

Supervised learning is a powerful tool businesses can use to gain a competitive edge in today's data-driven economy. It involves training an algorithm on a labeled dataset, where the correct output is known for each input. Supervised learning is used for tasks such as classification and regression and has many real-world applications in different sectors.

Unsupervised Learning

Unsupervised learning is a type of machine learning that involves training an algorithm on an unlabeled dataset where the correct output is not known. The algorithm learns to identify patterns in the data and group similar records together. Unsupervised learning is often used for tasks such as clustering and anomaly detection.

Clustering involves grouping records together based on their similarity. For example, an unsupervised learning algorithm can be used in marketing to group customers based on their purchasing behavior. This can help businesses to create more targeted marketing campaigns and improve customer engagement. In biology, an unsupervised learning algorithm can group genes together based on their expression patterns. This can help researchers to identify new drug targets and gain a deeper understanding of disease mechanisms.

Anomaly detection involves identifying records that are significantly different from the others. For example, in finance, an unsupervised learning algorithm can identify fraudulent transactions in a financial dataset. This can help businesses to prevent fraud and minimize losses. In manufacturing, an unsupervised learning algorithm can identify defective products in a production line. This can help companies to improve quality control and reduce waste.

Unsupervised learning algorithms are trained on an unlabeled dataset, typically divided into training and testing sets. The goal is to teach the algorithm to identify patterns in the data and group similar records together.

Unsupervised learning has many real-world applications in different sectors. For example, it can be used in finance to identify fraudulent transactions, in healthcare to group patients based on their medical histories, and in manufacturing to identify defective products.

Reinforcement Learning

Reinforcement learning is machine learning that enables an algorithm to learn by interacting with its environment. The algorithm learns to take actions that maximize a cumulative reward over time. Reinforcement learning is a type of learning driven by delayed feedback, meaning that the algorithm learns from the results of its actions after they are taken.

The reinforcement learning approach is modeled after how humans and animals learn from their environment. In essence, the algorithm learns by receiving feedback in the form of rewards or penalties for its actions and uses that feedback to adjust its future activities. This feedback loop is used to optimize the algorithm's behavior over time.

Reinforcement learning algorithms consist of an agent, a set of states, a collection of actions, and a reward function. The agent interacts with the environment by taking action, and the environment responds by providing the agent with a reward or penalty. The algorithm aims to learn a policy that maps each state to maximize the expected reward over time.

Reinforcement learning has many real-world applications, mainly when traditional programming approaches are not practical. For example, reinforcement learning has been successfully applied to problems such as game playing, robotics, and control systems. For instance, it has been used to train agents that can play games like chess and Go at superhuman levels, optimize the energy consumption of buildings, and develop robotic systems that can learn to navigate through complex environments.

One example of a real-world application of reinforcement learning is the development of AlphaGo, an artificial intelligence program developed by DeepMind. AlphaGo is a program that plays the board game Go, which is widely considered one of the most complex board games in the world. AlphaGo uses a combination of supervised learning and reinforcement learning to improve its performance over time. In 2016, AlphaGo defeated the world champion Go player, Lee Sedol, in a highly publicized match.

Another example of a real-world application of reinforcement learning is in robotics. Reinforcement learning has been used to develop robotic systems that can learn to navigate through complex environments, such as warehouses and manufacturing plants. These systems learn to optimize their movements and avoid obstacles by receiving feedback through rewards or penalties.

Reinforcement learning has also been used in healthcare to optimize treatment plans for patients based on their individual needs and responses to treatment. In finance, reinforcement learning has been used to optimize trading strategies and manage risk.

Neural Networks

Neural networks are machine learning algorithms modeled after the structure and function of the human brain. Neural networks consist of layers of interconnected nodes designed to recognize patterns in data. Each node in a neural network performs a simple mathematical operation on its inputs and passes the result to the next layer of nodes. The output of the last layer is the final prediction of the neural network.

Neural networks can be used for tasks such as image recognition, speech recognition, and natural language processing. In image recognition, a neural network can be trained to recognize specific objects in images, such as faces, buildings, or animals. In speech recognition, a neural network can remember spoken words and convert them to text. Finally, a neural network can generate human-like responses to text-based inputs in natural language processing.

Neural networks can be trained using supervised, unsupervised, or a combination of the two. In supervised learning, the network is trained on labeled data, where the correct output is known for each input. The network adjusts its weights and biases to minimize the difference between its predicted output and the correct output. In unsupervised learning, the network is trained on unlabeled data, where the correct output is unknown. The network adjusts its weights and biases to identify patterns in the data.

One of the most popular types of a neural networks is the convolutional neural network (CNN), which is particularly effective for image and video recognition tasks. CNNs use a combination of convolutional layers, which identify local patterns in the input, and pooling layers, which reduce the spatial dimensions of the information. The output of the last layer is typically passed through one or more fully connected layers, which perform a final classification or regression task.

Another popular type of neural network is the recurrent neural network (RNN), which is particularly effective for sequence data, such as time series or natural language data. RNNs use a feedback loop to allow information to persist across multiple time steps. This makes them well-suited for speech recognition, machine translation, and text generation tasks.

Neural networks have many real-world applications, particularly in computer vision, speech recognition, and natural language processing. For example, neural networks are used in autonomous vehicles to recognize road signs and traffic lights and in healthcare to diagnose medical images. They are also used in virtual assistants, such as Siri and Alexa, to identify and respond to natural language queries.

Neural networks are modeled after the structure and function of the human brain and can be used for tasks such as image recognition, speech recognition, and natural language processing. Neural networks can be trained using supervised, unsupervised, or a combination of the two. They have many real-world applications in different sectors, such as healthcare, finance, and retail.

Deep Learning

Deep learning is a machine learning subfield focused on training deep neural networks with many layers. Deep learning is particularly effective for tasks that involve large amounts of data and complex patterns, such as image recognition, speech recognition, and natural language processing. The key difference between deep learning and traditional machine learning is that deep learning algorithms can automatically learn feature representations from raw data rather than relying on hand-engineered features. As a result, deep learning algorithms are much more powerful and flexible than traditional machine learning algorithms.

Deep learning is based on the architecture of artificial neural networks, which are modeled after the structure and function of the human brain. The neural network consists of layers of interconnected nodes designed to recognize patterns in data. Each node in a neural network performs a simple mathematical operation on its inputs and passes the result to the next layer of nodes. The output of the last layer is the final prediction of the neural network.

Deep learning algorithms can be used for many tasks, including image recognition, speech recognition, natural language processing, and even playing games such as Go and chess. One of the most popular deep learning architectures is the convolutional neural network (CNN), which is particularly effective for image and video recognition tasks. CNNs use a combination of convolutional layers, which identify local patterns in the input, and pooling layers, which reduce the spatial dimensions of the input. The output of the last layer is typically passed through one or more fully connected layers, which perform a final classification or regression task. Another popular deep learning architecture is the recurrent neural network (RNN), which is particularly effective for sequence data, such as time series or natural language data. RNNs use a feedback loop to allow information to persist across multiple time steps. This makes them well-suited for speech recognition, machine translation, and text generation tasks.

Deep learning has many real-world applications, particularly in computer vision, speech recognition, and natural language processing. For example, deep learning is used in autonomous vehicles to recognize road signs and traffic lights and in healthcare to diagnose medical images. It is also used in virtual assistants, such as Siri and Alexa, to identify and respond to natural language queries.

Deep learning is a powerful tool that businesses can use to gain a competitive edge in today's data-driven economy. It is focused on training deep neural networks with many layers and is particularly effective for tasks that involve large amounts of data and complex patterns. Deep learning is based on the architecture of artificial neural networks, which are modeled after the structure and function of the human brain. As a result, it has many real-world applications in different sectors, such as healthcare, finance, and retail.

Real-World Applications of Machine Learning

Machine learning has many real-world applications across different industries and sectors. This section will provide examples of how machine learning improves efficiency, reduces costs, and creates new business opportunities.

Healthcare

In healthcare, machine learning is being used to improve patient outcomes and reduce costs. One of the primary applications of machine learning in healthcare is the analysis of medical images. Machine learning algorithms can be used to analyze images such as X-rays, MRIs, and CT scans and diagnose diseases such as cancer and heart disease. With the help of deep learning techniques, the accuracy of diagnosis can be improved. For example, Google has developed an algorithm that can detect breast cancer with a 99% accuracy rate.

Machine learning can also be used to develop personalized treatment plans for patients based on their individual needs and responses to treatment. Machine learning algorithms can analyze patient data and predict how well a patient will respond to a particular treatment. By providing personalized treatment plans, healthcare providers can improve patient outcomes and reduce the cost of care.

Finance

In finance, machine learning is used to optimize trading strategies and manage risk. For example, machine learning algorithms can be used to analyze market data and identify patterns that can be used to make more accurate predictions about market trends. This is particularly useful in algorithmic trading. Machine learning can also develop credit risk models and detect fraud. For example, machine learning algorithms can help identify potential fraud cases and prevent them by analyzing customer behavior and identifying patterns.

In addition, machine learning algorithms can be used for sentiment analysis in the financial industry. With natural language processing techniques, machine learning algorithms can analyze social media and news articles to identify trends in public opinion. These trends can help investors make better investment decisions and help financial institutions manage their portfolios.

Retail

In retail, machine learning improves the customer experience and increases sales. Machine learning algorithms can analyze customer data and provide personalized product recommendations. By providing relevant recommendations, retailers can improve the customer experience and increase sales.

Machine learning can also optimize inventory management and supply chain logistics. By analyzing historical sales data and predicting future demand, retailers can optimize their inventory management and reduce the cost of excess inventory. Additionally, machine learning algorithms can optimize supply chain logistics by predicting delivery times and identifying potential bottlenecks in the supply chain.

Manufacturing

Machine learning is used in manufacturing to improve operational efficiency and reduce costs. Machine learning algorithms can optimize production schedules and reduce downtime. By analyzing data from sensors on machines, machine learning algorithms can identify patterns that indicate when a machine is likely to break down. By predicting machine breakdowns, manufacturers can schedule preventative maintenance, reducing the cost of downtime and improving operational efficiency. Machine learning can also be used to detect defects in products. Machine learning algorithms can detect product defects in real-time by analyzing data from sensors on the production line. By detecting defects early in the production process, manufacturers can reduce waste and improve the quality of their products.

Marketing

In marketing, machine learning is used to improve targeting and increase the effectiveness of marketing campaigns. By analyzing customer data, machine learning algorithms can develop targeted marketing campaigns. For example, machine learning algorithms can analyze customer behavior and recommend products to customers most likely to buy them. This can improve the effectiveness of marketing campaigns and increase sales. Machine learning can also be used to predict customer churn. By analyzing customer behavior and identifying patterns that indicate a customer is likely to churn, machine learning algorithms can help businesses retain customers. Additionally, machine learning algorithms can identify customers most likely to purchase. By identifying high-value customers, businesses can focus their marketing efforts on them and increase sales.

Transportation

Machine learning is being used in transportation to improve safety and reduce costs. Machine learning algorithms can optimize routes and reduce fuel consumption. By analyzing traffic patterns and weather data, machine learning algorithms can predict the most efficient route for a driver. This can reduce fuel consumption and improve delivery times.

Machine learning can also be used to detect and prevent accidents. For example, machine learning algorithms can identify potential safety hazards and alert drivers in real-time by analyzing vehicle sensors' data. This can improve driver safety and reduce the cost of accidents.

Education

In education, machine learning is used to personalize learning and improve student outcomes. For example, machine learning algorithms can analyze student data and provide personalized learning plans for students. This can help students learn at their own pace and improve academic performance. Machine learning can also be used to improve the quality of education by analyzing data on teacher performance and student engagement. By identifying teacher performance and student engagement patterns, machine learning algorithms can help educators improve their teaching methods and identify areas where additional support is needed.

Energy

Machine learning improves power plants' efficiency and reduces energy production costs in the energy industry. For example, machine learning algorithms can analyze data from sensors in power plants and predict when equipment is likely to fail. As a result, machine learning algorithms can help power plant operators schedule maintenance and reduce downtime by predicting equipment failures.

Machine learning can also be used to optimize energy consumption. By analyzing data on energy usage and identifying patterns, machine learning algorithms can help businesses reduce their energy consumption and save money on their energy bills.

Machine learning has many real-world applications across different industries and sectors. It improves efficiency, reduces costs, and creates new business opportunities. By leveraging the power of machine learning, businesses can gain a competitive edge in today's data-driven economy. The examples above highlight just a few of how machine learning is being used to transform various industries. We expect to see even more innovative applications as machine learning technology advances.

III. Getting Started with Machine Learning

In the previous section, we discussed the different types of machine-learning algorithms and their real-world applications. In this section, we will focus on getting started with machine learning.

Getting started with machine learning can be a daunting task, especially for those who do not have a technical background. However, with the right approach and resources, anyone can learn how to use machine learning to solve business problems.

Selecting a Problem

The first step in getting started with machine learning is to select a problem to solve. This problem should be one that can be addressed with machine learning and is relevant to your business. Some examples of problems that can be addressed with machine learning include predicting customer churn, analyzing customer sentiment, and identifying fraud.

When selecting a problem, it is important to ensure that you have access to the data required to solve it. You will also need to ensure that the problem is well-defined and that you clearly understand the problem and the business context in which it exists.

Collecting and Preparing Data

The second step in machine learning is collecting and preparing data. This is a crucial step in the machine learning process, as the quality of the data directly impacts the performance of the machine learning model.

Data collection involves identifying and collecting data from various sources, including internal databases, publicly available data sources, and social media platforms. The data collected should be relevant to the problem being solved and sufficient in quantity and quality to support the machine learning process.

Once the data has been collected, it needs to be cleaned and preprocessed. This involves removing irrelevant or redundant data, filling in any missing values, and transforming the data into a format used by the machine learning algorithm. The data may also need to be normalized or standardized to ensure that it is on a consistent scale.

Identifying Relevant Data Sources

When identifying relevant data sources, it is important to consider the problem the machine learning model is trying to solve. The data sources should provide relevant information that is helpful for the model to make accurate predictions. Data can come from various sources, including internal databases, publicly available data sources, social media platforms, and third-party data providers.

For example, a credit card company might use customer transactional data to develop a machine-learning model to predict whether a particular transaction is fraudulent. In this case, the relevant data sources include transactional data, customer data, and other relevant data points.

Cleaning and Preprocessing the Data

The quality of the data used to train a machine-learning model can significantly impact the model's accuracy. To ensure that the data is of high quality, it is important to clean and preprocess the data before training the model.

Data cleaning involves removing irrelevant or redundant data, filling in missing values, and transforming the data into a format used by the machine learning algorithm. Preprocessing the data might involve techniques such as normalization, standardization, or feature scaling to ensure that the data is on a consistent scale.

For example, when training a model to predict the sentiment of customer reviews, it might be necessary to preprocess the data by removing stop words, converting text to lowercase, and applying stemming techniques to reduce word variation.

Ensuring Data Quality

Data quality is a critical factor in the performance of a machine learning model. Poor quality data can lead to incorrect predictions, lower accuracy, and suboptimal results. Therefore, ensuring that the data used to train a machine learning model is of high quality is important.

To ensure that the data is of high quality, it is important to consider the following factors:

- Accuracy: The data should be accurate and free of errors. Inaccurate data can lead to incorrect predictions and lower accuracy.

- Completeness: The data should be complete, and there should be no missing values. Missing values can lead to suboptimal results.

- Consistency: The data should be consistent and free of contradictions. Inconsistent data can lead to incorrect predictions.

- Relevance: The data should be relevant to the problem being solved. Irrelevant data can lead to lower accuracy.

- Timeliness: The data should be timely and up-to-date. Outdated data can lead to lower accuracy.

For example, when training a model to predict customer churn, it might be important to ensure that the data is accurate, complete, and relevant. Inaccurate data might lead to incorrect predictions, while incomplete data might lead to suboptimal results.

Selecting an Algorithm

The third step in machine learning is selecting an appropriate algorithm for the problem you are trying to solve. There are many different machine learning algorithms, each with strengths and weaknesses.

The most used machine learning algorithms include linear regression, logistic regression, decision trees, and neural networks. The selection of an algorithm will depend on the problem you are trying to solve and the characteristics of the data you have collected.

Evaluating the Model

The final step in machine learning is evaluating the model's performance. This involves testing the model on a testing dataset and comparing the predicted output to the actual output.

Many different metrics can be used to evaluate the performance of a machine learning model, including accuracy, precision, recall, and F1 score. The selection of a metric will depend on the problem you are trying to solve and the characteristics of the data you have collected.

Getting started with machine learning can be challenging, but by following the steps outlined in this section, anyone can learn how to use machine learning to solve business problems. By selecting a problem, collecting and preparing data, selecting an algorithm, and evaluating the model, businesses can leverage the power of machine learning to gain a competitive edge in today's data-driven economy. In the following pages of this section, we will discuss each of these steps in more detail.

Common Challenges in Implementing Machine Learning

In addition to the challenges discussed above, businesses face other common challenges when implementing machine learning. These include:

1. Overfitting: Overfitting occurs when a machine learning model is trained too well on the training data and performs poorly on new data. This can happen when the model is too complex or when there is not enough data to support the complexity of the model. To avoid overfitting, it is important to use techniques such as regularization or cross-validation to ensure that the model is not overly complex and to ensure that it generalizes well to new data.

2. Interpretability: In some cases, it may be important to understand

how a machine learning model makes its predictions. However, some machine learning algorithms, such as deep neural networks, can be difficult to interpret.

To address this challenge, researchers are developing new techniques for interpreting the output of machine learning models, such as layer-wise relevance propagation or feature visualization.

3. Ethics and Bias: Machine learning models can perpetuate bias if trained on biased data. This can lead to discrimination and inequities in decision-making. To address this challenge, it is important to carefully consider the data used to train the model and ensure that it is fair and unbiased.

4. Scalability: Machine learning models can require significant computational resources, and businesses must carefully consider the scalability of their machine learning infrastructure.

To address this challenge, businesses can consider cloud-based machine learning platforms that provide scalable and cost-effective infrastructure for machine learning.

5. Data Privacy and Security: Machine learning models often rely on sensitive data, such as personal or financial data. It is important to ensure that the data is properly secured and that privacy concerns are addressed.

To address this challenge, businesses can use differential privacy or secure multi-party computation techniques to protect sensitive data.

By addressing these challenges and following best practices for machine learning, businesses can successfully implement machine learning and gain a competitive advantage in today's data-driven economy.

IV. Implementing Machine Learning Algorithms

Steps Involved in Implementing Machine Learning

Machine learning algorithms have the potential to transform the way businesses operate, but implementing them can be a complex and challenging process. This section provides an overview of implementing a machine learning algorithm, from data preparation to model evaluation. The first step in implementing a machine learning algorithm is identifying the problem you want to solve. This may involve identifying a business problem that can be solved using machine learning or identifying a dataset that can be used to train a machine learning model.

Once you have identified the problem you want to solve, the next step is to collect and prepare the data. This may involve cleaning and preprocessing the data and selecting the features that will be used to train the machine learning model.

After the data has been prepared, the next step is to select an appropriate machine learning algorithm. This may involve choosing between different algorithms, such as supervised or unsupervised learning, and selecting an algorithm that suits the problem you are trying to solve.

Once an algorithm has been selected, the next step is to train the model using the prepared data. This may involve fine-tuning the model using hyperparameter tuning or cross-validation techniques.

Finally, the model is evaluated to determine its accuracy and performance. This may involve testing the model on a holdout dataset or using techniques such as A/B testing to compare the model's performance to other approaches.

In summary, implementing a machine learning algorithm involves several steps, from identifying the problem to evaluating the model's performance. By following best practices and addressing potential challenges, businesses can successfully implement machine learning and gain a competitive advantage.

Practical Guide to Machine Learning Implementation

Implementing machine learning in a business setting can be daunting, especially for those with limited technical expertise. This section provides a practical guide to implementing machine learning in a business setting, including tips for choosing the right algorithm and working with data scientists.

The first step in implementing machine learning in a business setting is identifying the business problem you want to solve. This may involve consulting different departments or stakeholders to determine where machine learning can impact the most.

Once a business problem has been identified, the next step is to gather the data required to train a machine learning model. This may involve collaborating with data scientists or experts to identify relevant datasets or collecting data from different sources.

The next step is to select an appropriate machine learning algorithm. This may involve working with data scientists to identify the best algorithm for the problem you are trying to solve or using online resources such as open-source libraries to find an algorithm well-suited to your needs.

After selecting an algorithm, the next step is to train the model using the prepared data. This may involve working with data scientists to fine-tune the model or using online resources to identify best practices for training machine learning models.

Finally, the model is evaluated to determine its accuracy and performance. This may involve working with data scientists to test the model on a holdout dataset or using online resources to learn how to evaluate the performance of machine learning models.

In summary, implementing machine learning in a business setting requires collaboration between different departments and experts and understanding the different steps involved in the implementation process. By following best practices and working with data scientists, businesses can successfully implement machine learning and gain a competitive advantage.

Potential Challenges and Solutions

Implementing machine learning in a business setting can be challenging, and businesses may face various potential challenges when implementing machine learning algorithms. In this section, we discuss some of the challenges businesses may face and provide solutions for addressing these challenges.

One of the biggest challenges businesses may face when implementing machine learning algorithms is the need for high-quality data. Machine learning algorithms rely on large, high-quality datasets to learn from. If the data is inaccurate, incomplete, or biased, the algorithm may not learn from it effectively. To address this challenge, businesses may need to invest in data cleaning and preprocessing or work with data scientists to develop data pipelines that ensure data quality and consistency.

Another challenge businesses may face when implementing machine learning is the need for interpretability and transparency in the model's decision-making process. For example, in some industries, such as healthcare and finance, it may be necessary to explain how the model arrived at a particular decision or recommendation. To address this challenge, businesses may need to use algorithms that provide interpretability, such as decision trees or linear models, or work with data scientists to develop methods for explaining the model's decision-making process.

Scalability is another challenge that businesses may face when implementing machine learning. Machine learning algorithms can be computationally intensive, and as datasets grow larger and more complex, it can become challenging to train and deploy models at scale. To address this challenge, businesses may need to invest in cloud infrastructure or work with data scientists to develop distributed computing solutions that can scale to handle larger datasets.

Finally, one of the biggest challenges businesses may face when implementing machine learning is the need for expertise and talent. Machine learning requires specialized knowledge and expertise, and businesses may struggle to find qualified data scientists and machine learning engineers. To address this challenge, businesses may need to invest in training and development programs or work with outside consultants or vendors to fill the expertise gap.

In summary, implementing machine learning algorithms can be challenging, and businesses may face various potential challenges when implementing machine learning. However, by addressing these challenges and following best practices, businesses can successfully implement machine learning and gain a competitive advantage.

Overcoming Data Quality Issues

Data quality is one of the most significant challenges businesses face when implementing machine learning. High-quality data is crucial for training machine learning models to make accurate predictions, and the quality of the data can significantly impact the model's accuracy and usefulness.

Data quality issues can arise from various sources, including incomplete or missing data, inconsistencies, errors, and bias. To address these challenges, businesses may need to invest in data cleaning and preprocessing or work with data scientists to develop data pipelines that ensure data quality and consistency.

One of the most important steps in ensuring data quality is identifying and addressing any missing or incomplete data. Data missing can occur for many reasons, such as technical errors, entry errors, or survey non-response. To address missing data, businesses may need to use imputation methods, such as mean imputation or hot-deck imputation, or work with data scientists to develop more advanced imputation methods, such as regression imputation or k-nearest neighbor imputation.

Another common data quality issue is inconsistencies and errors in the data. Inconsistencies can arise from differences in data formats, missing values, or encoding issues. To address these issues, businesses may need to standardize data formats or use data profiling and analysis tools to identify inconsistencies and errors in the data.

Finally, data quality issues can also arise from bias in the data. Bias can occur for many reasons, such as sample selection bias, measurement bias, or reporting bias. To address bias, businesses may need to work with data scientists to develop methods for detecting and correcting bias in the data, such as propensity score matching or calibration.

In summary, data quality is a critical component of machine learning. Businesses must ensure that the data used to train machine learning models is high quality and free from errors, inconsistencies, and bias. By addressing data quality issues, businesses can improve the accuracy and usefulness of machine learning models and gain a competitive advantage.

Understanding Algorithm Selection

Choosing the right machine learning algorithm for a given problem is crucial in implementing machine learning. Different machine learning algorithms have different strengths and weaknesses, and selecting the right algorithm can significantly impact the accuracy and usefulness of the model. There are several factors to consider when selecting a machine learning algorithm, including the nature of the problem being solved, the amount and type of available data, and the computational resources available.

One of the most important considerations when selecting a machine learning algorithm is the nature of the problem being solved. Some problems, such as classification or regression, are well-suited to supervised learning algorithms, while other problems, such as anomaly detection or clustering, may require unsupervised learning algorithms. Businesses may also need to consider whether the problem is linear or nonlinear and whether the data is structured or unstructured.

The amount and type of available data is also important consideration when selecting a machine learning algorithm. Some algorithms, such as decision trees or linear models, can work well with small datasets, while others, such as deep neural networks, may require large amounts of data to train effectively. Businesses may also need to consider the quality and consistency of the data, as well as the distribution of the data.

Finally, computational resources are important when selecting a machine learning algorithm. Some algorithms, such as deep neural networks, can be computationally intensive and require specialized hardware or distributed computing solutions to train effectively. Therefore, businesses may need to consider the computational resources available and choose an algorithm that can be trained within the available resources.

In summary, selecting the right machine learning algorithm is crucial in implementing machine learning. Businesses must consider several factors, including the nature of the problem being solved, the amount and type of available data, and the computational resources available. By selecting the right algorithm, businesses can

improve the accuracy and usefulness of machine learning models and gain a

competitive advantage.

Working with Data Scientists

Implementing machine learning algorithms can be a complex and

challenging process, and businesses may need to work with data scientists

to ensure the success of their machine learning projects. In this section, we

discuss some best practices for working with data scientists and ensuring

the success of machine learning projects.

The first step in working with data scientists is clearly defining the business

problem you are trying to solve. This may involve consulting different

departments or stakeholders to determine where machine learning can

impact the most. Clear communication is critical to ensure that data

scientists understand the business problem and can develop a machine

learning solution that meets your needs.

Once the business problem has been defined, the next step is to work with

data scientists to identify the data required to train a machine learning

model. Data scientists can help identify relevant datasets or collect data

from different sources. It is important to work with data scientists to ensure

that the data is high quality and suitable for training machine learning models.

The next step is to work with data scientists to select an appropriate machine learning algorithm. Data scientists can help identify the best algorithm for the problem you are trying to solve and ensure that the algorithm is well-suited to the data and the computational resources available.

After selecting an algorithm, the next step is to work with data scientists to train the model using the prepared data. This may involve fine-tuning the model using hyperparameter tuning or cross-validation techniques. Again, data scientists can help ensure that the model is trained effectively and that the results are accurate and reliable.

Finally, it is important to work with data scientists to evaluate the machine learning model's performance. This may involve testing the model on a holdout dataset or using techniques such as A/B testing to compare the model's performance to other approaches. Data scientists can help ensure that the model is evaluated effectively and that the results are interpretable and useful.

In summary, working with data scientists is critical to the success of machine learning projects. By following best practices and communicating with data scientists, businesses can ensure that machine learning solutions meet their needs and provide a competitive advantage.

Addressing Potential Challenges

Implementing machine learning algorithms can be challenging, and businesses may face various potential challenges when implementing machine learning. This section discusses some potential challenges businesses may face and provides solutions for addressing these challenges. One potential challenge is the need for high-quality data. Machine learning algorithms rely on large, high-quality datasets to learn from. If the data is inaccurate, incomplete, or biased, the algorithm may not learn from it effectively. To address this challenge, businesses may need to invest in data cleaning and preprocessing or work with data scientists to develop data pipelines that ensure data quality and consistency.

Another challenge that businesses may face is the need for interpretability and transparency in the model's decision-making process. For example, in some industries, such as healthcare and finance, it may be necessary to explain how the model arrived at a particular decision or recommendation. To address this challenge, businesses may need to use algorithms that provide interpretability, such as decision trees or linear models, or work with data scientists to develop methods for explaining the model's decision-making process.

Scalability is another challenge that businesses may face when implementing machine learning. Machine learning algorithms can be computationally intensive, and as datasets grow larger and more complex, it can become challenging to train and deploy models at scale. To address this challenge, businesses may need to invest in cloud infrastructure or work with data scientists to develop distributed computing solutions that can scale to handle larger datasets.

Finally, one of the biggest challenges businesses may face when implementing machine learning is

the need for expertise and talent. Machine learning requires specialized knowledge and expertise, and businesses may struggle to find qualified data scientists and machine learning engineers. To address this challenge, businesses may need to invest in training and development programs or work with outside consultants or vendors to fill the expertise gap.

In addition to these challenges, businesses may also need to consider legal and ethical considerations when implementing machine learning. For example, the use of machine learning algorithms in areas such as hiring, lending, and insurance may raise concerns about discrimination and bias. As a result, businesses may need to work with legal and compliance teams to ensure that machine learning models comply with relevant laws and regulations and do not create unintended consequences.

In summary, implementing machine learning algorithms can be challenging, and businesses may face various potential challenges when implementing machine learning. However, by addressing these challenges and following best practices, businesses can successfully implement machine learning and gain a competitive advantage.

Developing a Machine Learning Strategy

Developing a machine learning strategy is critical in ensuring machine learning projects' success. In this section, we discuss some best practices for developing a machine learning strategy and ensuring that machine learning projects meet the needs of the business.

The first step in developing a machine learning strategy is identifying the business problem or opportunity that machine learning can address. This may involve consulting different departments or stakeholders to determine where machine learning can impact the most. Clear communication is critical to ensure that machine learning solutions meet the needs of the business.

Once the business problem has been identified, the next step is to determine the scope of the machine learning project. This may involve identifying the required data, the algorithm used, and the available computational resources. Again, it is important to work with data scientists to ensure that the project's scope is realistic and achievable.

The next step is to develop a plan for data collection and preparation. This may involve identifying relevant datasets or collecting data from different sources. It is important to work with data scientists to ensure that the data is high quality and suitable for training machine learning models.

After the data has been collected and prepared, the next step is to select an appropriate machine learning algorithm. Again, data scientists can help identify the best algorithm for the problem you are trying to solve and ensure that the algorithm is well-suited to the data and the computational resources available.

Once an algorithm has been selected, the next step is to train the model using the prepared data. This may involve fine-tuning the model using hyperparameter tuning or cross-validation techniques. Again, data scientists can help ensure that the model is trained effectively and that the results are accurate and reliable.

Finally, it is important to evaluate the machine learning model's performance and ensure that it meets the needs of the business. This may involve testing the model on a holdout dataset or using techniques such as A/B testing to compare the model's performance to other approaches. Data scientists can help ensure that the model is evaluated effectively and that the results are interpretable and useful.

In summary, developing a machine learning strategy is a critical step in ensuring the success of machine learning projects. By following best practices and communicating with data scientists, businesses can ensure that machine learning solutions meet their needs and provide a competitive advantage.

Choosing the Right Algorithm

Choosing the right algorithm is critical in machine learning, as different algorithms may better suit different data types and problems. This section discusses some factors to consider when choosing a machine learning algorithm.

One of the first factors to consider is the type of problem you are trying to solve. For example, if you are working with text data, natural languages processing algorithms such as recurrent neural networks or convolutional neural networks may be a good choice. On the other hand, if you are working with structured data, algorithms such as decision trees or linear regression may be more appropriate.

Another factor to consider is the size and complexity of the dataset. For larger datasets, deep learning algorithms such as convolutional neural networks or recurrent neural networks may be more effective. In comparison, simpler algorithms such as decision trees or logistic regression may be more appropriate for smaller datasets.

The available computational resources are also important factors when choosing an algorithm. For example, deep learning algorithms can be computationally intensive and require specialized hardware such as GPUs or TPUs, while simpler algorithms may be more easily run on standard CPUs.

Interpretability is another factor to consider when choosing an algorithm, particularly in industries where transparency and interpretability are important. For example, decision trees or linear models are often more interpretable than deep learning algorithms such as neural networks.

Finally, it is important to consider the availability of pre-trained models and libraries when choosing an algorithm. Pre-trained models and libraries can save time and resources when developing machine learning models, particularly for common tasks such as image recognition or natural language processing.

By considering these factors, businesses can choose the machine learning algorithm that is best suited to their needs and ensure the success of their machine learning projects.

Data Preparation

Data preparation is a critical step in implementing machine learning, as the data's quality and suitability can significantly impact the machine learning model's performance. In this section, we discuss some best practices for data preparation.

The first step in data preparation is identifying the data required to train a machine learning model. This may involve identifying relevant datasets or collecting data from different sources. It is important to work with data scientists to ensure that the data is high quality and suitable for training machine learning models.

Once the data has been collected, the next step is to clean and preprocess the data. This may involve removing duplicate or irrelevant data, converting data to a consistent format, and handling missing or incomplete data. Again, it is important to work with data scientists to ensure that the data is cleaned and preprocessed effectively and that the resulting dataset is suitable for training machine learning models.

The next step is to split the data into training and testing datasets. The training dataset is used to train the machine learning model, while the testing dataset is used to evaluate the model's performance. Again, it is important to work with data scientists to ensure that the data is split effectively and that the resulting datasets represent the data the model will use.

Finally, it is important to consider the balance of the data when preparing it for machine learning. In some cases, the data may be imbalanced, with one class or outcome being more common than others. This can lead to biased or inaccurate models. To address this, businesses may need to use techniques such as oversampling or undersampling to balance the data.

In summary, data preparation is a critical step in implementing machine learning, and businesses must ensure that their data is of high quality and suitable for training machine learning models. By following best practices and working with data scientists, businesses can prepare their data effectively and ensure the success of their machine learning projects.

Training and Testing the Model

Once the data has been prepared, the next step is to train and test the machine learning model. This section discusses some best practices for training and testing machine learning models.

The first step in training a machine learning model is to select an appropriate algorithm and hyperparameters. This may involve experimenting with different algorithms and hyperparameters to find the best combination for the problem at hand. Data scientists can help identify the best algorithm and hyperparameters for the problem.

Once an algorithm and hyperparameters have been selected, the next step is to train the model using the prepared data. This may involve fine-tuning the model using hyperparameter tuning or cross-validation techniques. Again, it is important to work with data scientists to ensure that the model is trained effectively and that the results are accurate and reliable.

After the model has been trained, the next step is to test the model using a holdout dataset. This can help evaluate the model's performance and identify any areas for improvement. In addition, data scientists can help evaluate the model's performance and identify any issues that need to be addressed.

It is also important to evaluate the model's performance on real-world data. This may involve deploying the model in a production environment and monitoring its performance over time. Data scientists can help ensure that the model performs effectively and that any issues are addressed promptly. In summary, training and testing the machine learning model is critical in implementing machine learning. By following best practices and working with data scientists, businesses can ensure that their models are trained effectively and perform accurately in real-world environments.

Model Deployment and Monitoring

Once the machine learning model has been trained and tested, the next step is to deploy the model in a production environment. This section discusses some best practices for model deployment and monitoring.

The first step in deploying a machine learning model is to integrate the model into the production environment. This may involve developing APIs or interfaces that allow the model to be accessed by other applications or systems. Data scientists can help ensure that the model is integrated effectively and accessible to the appropriate stakeholders.

Once the model has been integrated, it is important to monitor its performance over time. This can help identify any issues or changes in performance and allow for prompt resolution. In addition, data scientists can help monitor the model's performance and identify any issues that need to be addressed.

It is also important to ensure that the model is updated and maintained over time. This may involve retraining the model on new data or fine-tuning it to improve its performance. Again, data scientists can help ensure that the model is updated and maintained effectively over time.

Finally, it is important to ensure that the model is secure and that data privacy and confidentiality are maintained. Businesses may need to work with cybersecurity and compliance teams to ensure that the model is secure and compliant with relevant laws and regulations.

In summary, deploying and monitoring a machine learning model is critical in implementing machine learning. By following best practices and working with data scientists, businesses can ensure that their models are deployed effectively and maintained over time.

Measuring Success

Measuring the success of a machine learning project is important to ensure that the project has met the goals of the business. In this section, we discuss some metrics that can be used to measure the success of a machine learning project.

One of the most common metrics for measuring the success of a machine learning project is accuracy. Accuracy measures the percentage of correct predictions the model makes on a given dataset. However, accuracy may not always be the best metric to use, as it does not account for the cost of false positives or false negatives.

Another metric to consider is precision and recall. Precision measures the proportion of true positives among all positive predictions, while recall measures the proportion of true positives among all actual positive cases. These metrics can be particularly useful in cases where the cost of false positives or negatives is high.

In addition to these metrics, businesses may also need to consider other factors when measuring the success of a machine learning project, such as speed, scalability, and interpretability. Therefore, it is important to work with data scientists and other stakeholders to identify the most appropriate metrics for the project and ensure that they are measured consistently over time.

By measuring the success of machine learning projects, businesses can identify areas for improvement and ensure that their projects meet the business's goals.

Best Practices for Machine Learning Implementation

This section summarizes some best practices for implementing machine learning in a business setting.

The first best practice is identifying the business problem or opportunity that machine learning can address. Then, clear communication is critical to ensure that machine learning solutions meet the needs of the business.

The second best practice is to work with data scientists and other stakeholders to choose the right algorithm and prepare the data effectively. By following best practices for data preparation and algorithm selection, businesses can ensure that their machine learning models are accurate and reliable.

The third best practice is measuring machine learning projects' success using appropriate metrics. By measuring success, businesses can identify areas for improvement and ensure that their machine learning projects are meeting the business's goals.

The fourth best practice is continuously monitoring and updating machine learning models over time. By monitoring performance and updating models, businesses can ensure that their machine learning solutions are effective and provide a competitive advantage.

By following these best practices, businesses can successfully implement machine learning and gain a competitive advantage.

Potential Challenges and Solutions

While implementing machine learning can bring significant benefits to businesses, potential challenges must be addressed. This section discusses some challenges businesses may face when implementing machine learning and provides solutions for addressing these challenges.

One of the main challenges that businesses may face is a lack of data or poor data quality. This can impact the accuracy and reliability of machine learning models. To address this, businesses may need to collect additional data or work with data scientists to clean and preprocess the data effectively.

Another challenge is a lack of technical expertise. Machine learning requires specialized skills, and businesses may not have the expertise in-house to implement machine learning effectively. To address this, businesses may need to hire data scientists or work with third-party providers to develop and implement machine learning solutions.

A lack of transparency or interpretability can also be a challenge, particularly in industries where transparency and accountability are important. Deep learning algorithms, for example, can be difficult to interpret, which can be a barrier to adoption. To address this, businesses may need to prioritize using simpler algorithms or work with data scientists to develop methods for interpreting complex models.

Finally, there may be challenges related to data privacy and security. Machine learning models often rely on sensitive data, and it is important to ensure that this data is protected from unauthorized access or misuse. Businesses may need to work with cybersecurity and compliance teams to ensure that their machine-learning solutions are secure and compliant with relevant laws and regulations.

Businesses can implement machine learning effectively by addressing these challenges and gaining a competitive advantage.

Conclusion

This book has provided an overview of machine learning and its applications in a business setting. We have discussed the different types of machine learning algorithms, their real-world applications, and best practices for implementing machine learning and addressing potential challenges.

Implementing machine learning effectively, businesses can gain a competitive advantage, improve operational efficiency, and achieve better business outcomes. But, it is important to follow best practices and work with data scientists and other stakeholders to ensure that machine learning solutions are accurate, reliable, and value the business.

We hope this book has comprehensively introduced machine learning and its potential benefits for businesses. We encourage readers to explore and continue learning about machine learning and its applications in their industry.

References

This section provides a list of references used throughout the book. These references include articles, books, and other resources that can provide further information on machine learning and its applications in business.

- Alpaydin, E. (2010). Introduction to machine learning (2nd ed.). Cambridge, MA: MIT Press.

- Goodfellow, I., Bengio, Y., & Courville, A. (2016). Deep learning. Cambridge, MA: MIT Press.

- Hastie, T., Tibshirani, R., & Friedman, J. (2009). The elements of statistical learning: Data mining, inference, and prediction (2nd ed.). New York: Springer.

- Jordan, M. I., & Mitchell, T. M. (2015). Machine learning: Trends, perspectives, and prospects. Science, 349(6245), 255-260.

- Kelleher, J. D., Tierney, B., & Tierney, B. (2018). Data science: An introduction (2nd ed.). Boca Raton, FL: CRC Press.

- Murphy, K. P. (2012). Machine learning: A probabilistic perspective. Cambridge, MA: MIT Press.

- Ng, A. (2017). Machine learning yearning. Technical report, Andrew Ng.

- Provost, F., & Fawcett, T. (2013). Data science for business: What you need to know about data mining and data-analytic thinking.

Sebastopol, CA: O'Reilly Media.

These references provide a range of perspectives and resources for readers who want to learn more about machine learning and its applications in business. By exploring these resources, readers can gain a deeper understanding of the topic and explore different approaches to implementing machine learning in their organizations.

V. Applications of Machine Learning in Business

Machine learning has become an essential technology in today's data-driven economy. Businesses that can leverage machine learning have the potential to gain a competitive edge by improving their operations, enhancing their products and services, and improving customer engagement. In this section, we will explore the different sectors of the economy that can benefit from machine learning and provide real-world examples of how machine learning is being used in business. We will also discuss how machine learning can improve operational efficiency and drive growth in a business setting. The widespread use of technology has led to a massive increase in the volume of data available to businesses. However, the ability to transform this data into insights and actionable outcomes are only possible through machine learning. Therefore, machine learning algorithms have been developed to detect patterns in data, enabling businesses to make accurate predictions and informed decisions.

It is important to note that implementing machine learning is not a one-size-fits-all approach. The choice of algorithm, data quality, and expertise of the data scientists all affect the success of machine learning implementation. In this section, we will explore the different applications of machine learning in business and guide how to effectively integrate machine learning into business operations.

Sectors of the Economy that can Benefit from Machine Learning

In this section, we will explore the different sectors of the economy that can benefit from machine learning. We will discuss the specific applications of machine learning in healthcare, finance, retail, manufacturing, transportation, and energy sectors.

Healthcare: The healthcare sector is one of the most promising areas for the application of machine learning. With the vast amount of medical data available, machine learning can be used to analyze patient data and improve diagnosis and treatment. For example, machine learning can be used to analyze medical images and identify patterns that are not visible to the human eye, aiding in diagnosing diseases like cancer. Additionally, machine learning can help identify patients at high risk of developing certain diseases and predict patient outcomes, improving the overall quality of care.

Finance: The finance sector can benefit greatly from machine learning for fraud detection, risk assessment, and credit scoring. Machine learning algorithms can detect fraudulent activity and prevent unauthorized purchases by analyzing transactional data and identifying patterns.

Additionally, machine learning can predict market trends and optimize investment strategies, improving the bottom line for businesses in the financial industry.

Retail: Machine learning can be used for personalized marketing, inventory management, and demand forecasting in the retail sector. By analyzing customer data and identifying patterns in purchasing behavior, businesses can create targeted marketing campaigns that increase customer engagement and drive sales. Additionally, machine learning can help to optimize inventory levels, reducing costs and improving efficiency in the supply chain.

Manufacturing: In the manufacturing sector, machine learning can be used for predictive maintenance, quality control, and supply chain optimization. By analyzing sensor data, machine learning algorithms can predict when equipment will likely fail, enabling businesses to schedule maintenance before a breakdown occurs. Additionally, machine learning can ensure consistent product quality by identifying anomalies and deviations in the manufacturing process.

Transportation: Machine learning can be used for route optimization, demand forecasting, and predictive maintenance in the transportation sector. Machine learning algorithms can identify the most efficient routes by analyzing traffic patterns and historical data, reducing travel time and costs.

Additionally, machine learning can predict demand and optimize the use of transportation assets.

Energy: In the energy sector, machine learning can be used for predictive maintenance and energy consumption optimization. By analyzing sensor data from energy infrastructure, machine learning algorithms can predict equipment failures and schedule maintenance before a breakdown occurs. Additionally, machine learning can optimize energy consumption by analyzing energy usage patterns and identifying areas for improvement.

Machine learning can significantly benefit businesses in a wide range of sectors. By analyzing data and identifying patterns, businesses can gain previously impossible insights and make informed decisions that can drive growth and improve operational efficiency.

Real-World Applications of Machine Learning

This section will provide examples of real-world applications of machine learning in business. These examples highlight the impact machine learning can have on a business's bottom line and demonstrate the potential of machine learning to transform business operations.

Personalized Marketing: Machine learning can be used to analyze customer data and provide personalized recommendations, improving customer engagement and increasing sales. For example, Netflix uses machine learning to analyze viewer data and provide recommendations for new shows and movies based on past viewing history.

Predictive Maintenance: Machine learning can predict when equipment will likely fail, enabling businesses to schedule maintenance before a breakdown occurs. This can help to reduce downtime and save money on repair costs. For example, General Electric uses machine learning to predict when jet engines are likely to fail, allowing them to schedule maintenance before a breakdown occurs.

Fraud Detection: Machine learning can detect fraudulent activity in real-time, preventing unauthorized purchases and reducing the risk of financial loss. For example, PayPal uses machine learning to detect fraudulent activity on its platform, allowing them to prevent unauthorized transactions before they occur.

Demand Forecasting: Machine learning can be used to predict demand for products and services, allowing businesses to optimize inventory levels and reduce costs. For example, Amazon uses machine learning to predict product demand and optimize inventory levels, reducing costs and improving efficiency.

Natural Language Processing: Machine learning can be used for natural language processing, enabling businesses to analyze large amounts of text data and extract insights. For example, IBM's Watson platform uses machine learning to analyze text data and provide insights into customer sentiment and market trends.

Image and Speech Recognition: Machine learning can be used for image and speech recognition, enabling businesses to extract data from visual and audio sources. For example, Google uses machine learning to provide image and speech recognition services, allowing businesses to extract data from images and audio files.

Credit Scoring: Machine learning can analyze credit data and provide accurate credit scores, enabling businesses to make informed lending decisions. For example, ZestFinance uses machine learning to analyze credit data and provide more accurate credit scores, improving the accuracy of lending decisions.

Medical Diagnosis: Machine learning can analyze medical data and improve diagnosis and treatment. For example, IBM's Watson platform analyzes medical data and provides more accurate cancer diagnoses.

Improving Operational Efficiency and Driving Growth

Improving operational efficiency is one of the key benefits of machine learning. By analyzing data and identifying patterns, businesses can gain insights that help them to optimize their operations and reduce costs. This section will discuss how machine learning can improve operational efficiency and drive growth in a business setting.

Supply Chain Optimization: Machine learning can be used to optimize the supply chain, reducing costs and improving efficiency. By analyzing historical data and identifying patterns, businesses can optimize their inventory levels, reduce transportation costs, and improve the overall efficiency of the supply chain. For example, Walmart uses machine learning to optimize its supply chain, reducing transportation costs by over $100 million.

Process Automation: Machine learning can automate repetitive and time-consuming tasks, reducing employee workload and improving the overall efficiency of business operations. For example, JPMorgan uses machine learning to automate repetitive legal tasks, reducing the time it takes to review legal documents by over 90%.

Customer Engagement: Machine learning can improve customer engagement, providing personalized recommendations and improving the overall customer experience. By analyzing customer data and identifying patterns in purchasing behavior, businesses can provide targeted recommendations that increase customer engagement and drive sales. For example, Amazon uses machine learning to provide personalized customer recommendations, improving customer engagement and increasing sales.

Predictive Analytics: Machine learning can provide predictive analytics, enabling businesses to make informed decisions and take action before problems occur. Businesses can predict equipment failures, identify customer churn, and optimize operations by analyzing data and identifying patterns. For example, Uber uses machine learning to predict rider demand and optimize driver availability, improving the overall efficiency of their operations.

Conclusion

Machine learning has become an essential technology in today's data-driven economy. By analyzing data and identifying patterns, businesses can gain insights that help them to optimize their operations and reduce costs. The potential applications of machine learning in business are vast, and the benefits can be significant.

However, it is important to note that implementing machine learning is not a one-size-fits-all approach. The choice of algorithm, data quality, and expertise of the data scientists all affect the success of machine learning implementation. Businesses must carefully consider these factors and work with data scientists to develop effective machine learning strategies.

In this section, we have explored the different sectors of the economy that can benefit from machine learning and provided real-world examples of how machine learning is being used in business. We have also discussed how machine learning can improve operational efficiency and drive growth in a business setting.

As technology evolves, machine learning applications will likely become even more widespread. Therefore, businesses need to stay up to date with the latest advancements in machine learning and consider how they can be used to improve their operations and gain a competitive edge in their respective markets.

VI. Conclusions

Throughout this book, we have explored the concept of machine learning and its potential applications in business. We have defined key terms and concepts, provided an overview of the most common machine learning algorithms used in business, and discussed how to get started with machine learning.

In this concluding section, we will summarize the key takeaways from the book and provide some final thoughts and recommendations for readers interested in implementing machine learning in their businesses.

Key Takeaways

1. Machine learning is a powerful technology that can provide businesses with a competitive edge by analyzing data and identifying patterns.

2. Understanding the key concepts and algorithms used in machine learning is essential for businesses that want to leverage this technology.

3. The sectors of the economy that can benefit from machine learning are diverse and include healthcare, finance, and retail.

4. Implementing machine learning is not a one-size-fits-all approach, and businesses must carefully consider the choice of algorithm, data quality, and the expertise of the data scientists involved.

5. Improving operational efficiency and driving growth are two key benefits of machine learning, and businesses should consider how they can leverage this technology to optimize their operations.

Final Thoughts and Recommendations

As we wrap up this book, it is important to note that machine learning is a rapidly evolving field, and businesses that want to stay ahead of the curve must be willing to adapt and innovate.

For businesses that are interested in implementing machine learning, we recommend the following steps:

1. Start with a clear understanding of your business objectives and the data you have available.

2. Choose the right algorithm for your business needs and data set.

3. Invest in the expertise of data scientists and other professionals who can help you implement machine learning effectively.

4. Continuously evaluate and refine your machine learning models to ensure that they are delivering optimal results.

5. Keep up with the latest machine learning advancements and

explore new applications for this technology in your business. By following these recommendations, businesses can leverage the power of machine learning to gain a competitive edge and optimize their operations.

Looking to the Future

As we look to the future of machine learning, it is clear that this technology will continue to play an increasingly important role in business operations. The potential applications of machine learning are vast, and we are just beginning to scratch the surface of what is possible.

In the coming years, we can expect machine learning to become even more widespread, with more businesses leveraging this technology to gain insights and optimize their operations.

Businesses need to stay updated with the latest advancements in machine learning and explore new applications for this technology. By embracing machine learning and investing in the right talent and technology, businesses can position themselves for success in the increasingly data-driven economy of the future.

In conclusion, we hope this book has provided a valuable introduction to machine learning and its potential applications in business. We have covered key concepts and algorithms, provided real-world examples of how machine learning is used in various sectors, and discussed how to implement machine learning in a business setting.

We encourage readers to continue exploring machine learning and embrace this powerful technology as a tool for gaining a competitive edge and optimizing their operations. By staying up to date with the latest advancements in machine learning and continuing to innovate, businesses can position themselves for success in the data-driven economy of the future.

VII. Appendices

Appendix A: Tools and Frameworks

This appendix overviews some of the most popular tools and frameworks used in machine learning, including programming languages, libraries, and development environments.

By exploring these tools and frameworks, readers can better understand the technical aspects of machine learning and the resources available for developing and implementing machine learning solutions.

Appendix B: Data Preparation Checklist

This appendix provides a checklist of best practices for data preparation in machine learning, including data cleaning, preprocessing, and feature selection. By following this checklist, readers can ensure that their data is prepared effectively for use in machine learning models.

Appendix C: Algorithm Selection Guide

This appendix provides a guide to selecting the right algorithm for a given machine learning task, including considerations such as the data type, the dataset size, and the specific business problem being addressed.

Appendix D: Glossary

This appendix provides a more comprehensive glossary of key terms related to machine learning, including technical terms and acronyms.

By including these appendices, we provide additional resources and information to help readers explore machine learning further and implement machine learning solutions effectively in their organizations.

Appendix A - Tools and Frameworks

This appendix overviews some of the most popular tools and frameworks used in machine learning, including programming languages, libraries, and development environments.

Programming Languages:

- Python: A popular language for machine learning, with a large and active community, as well as numerous libraries and tools for data science and machine learning.

- R: A language and environment for statistical computing and graphics, with various packages and libraries for data science and machine learning.

- Java is a versatile language used for many applications, including machine learning.

Libraries and Frameworks:

- Scikit-learn: A popular Python library for machine learning with tools for classification, regression, clustering, and more.

- TensorFlow: A flexible and powerful library for deep learning developed by Google.

- PyTorch: A deep learning library with a dynamic computation graph developed by Facebook.

- Keras: A high-level API for building and training deep learning models, with support for multiple backends, including TensorFlow and Theano.

- Apache Spark: A fast and powerful big data processing engine with support for machine learning and other advanced analytics.

Development Environments:

- Jupyter Notebook: A web-based interactive environment for data science and machine learning, with support for multiple languages, including Python and R.

- Spyder: An integrated development environment (IDE) for scientific computing and data science, with features for editing, debugging, and visualization.

- RStudio: An IDE for R programming with data science and machine learning features.

By exploring these tools and frameworks, readers can better understand the technical aspects of machine learning and the resources available for developing and implementing machine learning solutions.

Appendix B - Data Preparation Checklist

In this appendix, we provide a checklist of best practices for data preparation in machine learning, including data cleaning, preprocessing, and feature selection.

1. Identify and handle missing data:

- Identify missing values in the dataset

- Decide how to handle missing data (e.g., imputation, deletion)

2. Identify and handle outliers:

- Identify outliers in the dataset

- Decide how to handle outliers (e.g., removal, transformation)

3. Feature selection:

- Identify the relevant features for the machine learning task

- Remove irrelevant or redundant features

4. Data normalization:

- Normalize data to ensure that features have a similar scale

- Normalize data using techniques such as z-score normalization or min-max normalization

5. Data encoding:

- Convert categorical data to numerical data using techniques such as one-hot encoding or label encoding

- Convert text data to numerical data using techniques such as bag-of-words or TF-IDF

6. Train-test split:

- Split the dataset into a training set and a test set

- Use the training set to train the machine learning model and the test set to evaluate the model's performance

By following this checklist, readers can ensure that their data is prepared effectively for use in machine learning models.

Appendix C - Algorithm Selection Guide

In this appendix, we provide a guide to selecting the right algorithm for a given machine learning task, including considerations such as the data type, the dataset size, and the specific business problem being addressed.

1. Classification:

- Consider the type of data (e.g., categorical, numerical)

- Consider the size of the dataset (e.g., small, large)

- Consider the specific business problem being addressed (e.g., fraud detection, sentiment analysis)

- Common algorithms include logistic regression, decision trees, and support vector machines

2. Regression:

- Consider the type of data (e.g., continuous, categorical)

- Consider the size of the dataset (e.g., small, large)

- Consider the specific business problem being addressed (e.g., predicting sales, forecasting demand)

- Common algorithms include linear regression, polynomial regression, and decision trees

3. Clustering:

- Consider the type of data (e.g., numerical, categorical)

- Consider the size of the dataset (e.g., small, large)

- Consider the specific business problem being addressed (e.g., customer segmentation, anomaly detection)

- Common algorithms include k-means clustering, hierarchical clustering, and DBSCAN

4. Dimensionality Reduction:

- Consider the size of the dataset (e.g., large, very large)

- Consider the specific business problem being addressed (e.g., data visualization, feature extraction)

- Common algorithms include principal component analysis (PCA), t-SNE, and autoencoders

By considering these factors and selecting the appropriate algorithm for a given machine learning task, readers can ensure that they are using the most effective and efficient approach to solving their business problems.

Appendix D - Glossary

In this appendix, we provide a more comprehensive glossary of key terms related to machine learning, including technical terms and acronyms.

- Artificial Intelligence (AI): A field of study focused on creating machines that can perform tasks that would typically require human intelligence, such as natural language processing, image recognition, and decision-making.

- Big Data: A term used to describe large and complex datasets that are too difficult to process using traditional data processing methods.

- Deep Learning: A subfield of machine learning focused on neural networks with multiple layers for tasks such as image recognition and natural language processing.

- Feature Engineering: The process of selecting and transforming input data to create new features that are more relevant or useful for a machine learning task.

- Machine Learning (ML): A field of study focused on creating algorithms that can learn from data and make predictions or decisions based on that learning.

- Neural Network: A machine learning algorithm modeled after the structure of the human brain, used for tasks such as image recognition and natural language processing.

- Overfitting: A situation in which a machine learning model is too complex and becomes too specific to the training data, resulting in poor performance on new, unseen data.

- Regression: A machine learning task focused on predicting a continuous output variable, such as a house's price or a city's temperature.

- Supervised Learning: A type of machine learning in which the algorithm is trained using labeled data to predict a specific output based on the input data.

- Unsupervised Learning: A type of machine learning in which the algorithm is trained using unlabeled data to identify patterns or relationships in the data.

- Algorithm: A set of instructions for solving a problem or performing a task.

- Supervised Learning: A type of machine learning in which an algorithm is trained on a labeled dataset, where the correct output is known for each input.

- Unsupervised Learning: A type of machine learning in which an algorithm is trained on an unlabeled dataset, where the correct output is unknown.

- Reinforcement Learning: A machine learning in which an algorithm learns to take actions in an environment to maximize a reward.

- Data Cleaning: The process of removing errors and inconsistencies from data.

- Data Normalization: Transforming data to be comparable across different variables.

- Feature Engineering: The process of extracting meaningful features from data that can be used to train a machine learning algorithm.

- Model Evaluation: The process of assessing the performance of a machine learning model using metrics such as accuracy and precision.

- Precision: The proportion of true positives out of all positive predictions a machine learning model makes.

- Accuracy: The proportion of correct predictions made by a machine learning model.

By using this glossary, readers can gain a deeper understanding of the technical terms and concepts related to machine learning, helping them to communicate more effectively with data scientists and other technical experts in their organizations.

ABOUT THE AUTHOR

Andrei Oprisan spent the last 15+ years helping companies grow using technology.

Andrei is an alumnus of Columbia University with an MS in computer science, machine learning track. He has managed large teams and led transformations and organizational turnarounds from start-ups to large companies in eCommerce, marketing, sales, and advertising at scale at companies like HubSpot, Liberty Mutual, Wayfair, and OneScreen.AI. He has led data science teams and implemented patent-pending machine learning algorithms for geospatial analytics and audience prediction for sales and marketing campaign optimization.

In addition to technical expertise, Andrei is skilled in working with business stakeholders to identify and address business problems and opportunities using machine learning. He has worked with various organizations, from startups to Fortune 500 companies, and has a deep understanding of the challenges and opportunities involved in implementing machine learning in a business setting.

Drawing on his extensive experience and expertise, Andrei has developed this book as a comprehensive introduction to machine learning for a general business audience. He hopes readers will find the book informative, engaging, and useful in their efforts to implement machine learning and gain a competitive advantage in their industry.